A Cotman Colour Book
with text by **Robin Smith**

Life of the Ponds and Streams in Britain Book 1

Jarrold Colour Publications, Norwich

It never ceases to amaze me how so many people interested in fishing, rivers, lakes and ponds are solely concerned with the fish life and know so little of the complex ecological factors which allow these vertebrates to survive. However, the fascinating subject of ecology would fill several hundred volumes of this size, but I hope that this brief guide to the freshwater invertebrate life of the British Isles will serve to explain some of the complex food chains, population explosions and amazing events which can occur in, and around, a body of fresh water.

The basic idea of an ecologically balanced environment starts with energy from the sun, in the form of light, enabling plants to convert minerals, water and carbon dioxide into plant material by a process known as photosynthesis. These plants, either as living or dead material, are then eaten by animals which are, in turn, eaten by larger animals, and so on until, as these animals and plants die, bacteria take over, breaking down the complex proteins and carbohydrates into their basic elements – minerals and carbon dioxide – and so the cycle continues.

I should, of course, point out that this is a very much simplified picture of the ideal situation. In reality so many factors can affect this balance that very often an environment can undergo severe stress and consequent changes. Pollution, in its various forms, can have drastic effects on the animal life in any particular environment but especially in water.

In this first book I have tried to include some of the more common invertebrate animals which play important parts in the maintenance of an ecologically balanced environment, and which spend most of their lives in water. The second book deals with the larval, pupal and adult stages of those insects which have terrestrial adults such as the gnats, mayflies, dragon flies and caddis flies, etc. These too have very important roles in the balanced ecology but, unfortunately, because of the numbers of species involved it was not possible to include both types in the same volume.

Most of these animals can be found quite easily, and with the help of a small pond net and a jam jar an interesting collection can be built up in a very short time. The problem then arises as to how to identify the animals one has caught. To the experienced eye a quick glance will enable an approximate identification but often there are so many similar species that it is impossible, without very close examination, to name the animal accurately. However, this is by no means essential to the understanding of the animal's place in the ecological system; for example, to say that something is a water boatman *Corixa* species is functionally accurate since all species of the genus *Corixa* have basically the same appearance, habits and life style. Only the specialist who requires to know the exact species needs to delve into the minute anatomical features which identify the individual concerned. On the other hand, to say that an animal is a beetle is perhaps too vague, since the variety of beetles found in Britain extends from minute crawling beetles, feeding on dead plant material, to the large swimming varieties catching and eating living animals. With this in mind I have attempted, not so much to identify accurately, but to explain the ecological position which the animal holds and its function in the maintenance of the environment in which it lives.

The diversity of the animal life found in any particular environment depends on several factors, especially in aquatic situations. In general, different species will be found in running water from those found in still water but there are several animals which are equally adapted to survive in both types. Fast-flowing streams and rivers tend to hold more of those species which require a high oxygen level, whereas still water and especially stagnant ponds will hold only those species capable of surviving in low oxygen levels. Those which feed on plants, or live in mud will, usually, only be found in such surroundings as suit their needs. Thus it is possible to build up a picture of what sort of animals are likely to be found in a given body of water.

In most areas, a basic stock of the more common animals can be found regardless of the type of water. These include the freshwater shrimp *Gammarus*, the water boatman *Corixa*, the water fleas *Daphnia* and *Cyclops*, flatworms *Polycelis* and *Dendrocoelium* and several of the larval stages of species mentioned in the second book. Those animals found in addition to these can often indicate the type of water being studied. A very good example is the crayfish, which occurs almost solely in fast running water with a high oxygen level and usually a high calcium level.

There are, of course, exceptions to these rules, particularly where waters have been stocked with species not usually found in that particular type of environment. (Crayfish have been found in deep muddy lakes but only in very small numbers.)

Other species are restricted to certain areas in Britain, simply because they have no means of covering large distances outside their aquatic environment and this applies particularly to those species which cannot fly. It is easily forgotten that many aquatic insects can fly, a very good example being *Corixa*, the water boatman. This is one of the first animals to populate a new body of water, usually appearing within a matter of days after a reasonable surface area has been formed. Other animals such as beetles and the larvae of many terrestrial insects soon follow, and very quickly a wide variety of animal life is built up. Those species incapable of flight arrive on the scene as spores, or perhaps as eggs attached to birds' legs or even as parasitic stages attached to some of the flying insects. Indeed, the opportunity of studying the animals arriving at a new site should not be missed, you will be amazed at how quickly a very diverse community of invertebrates is built up.

As the various populations of animals and plants arrive, those best suited to the particular conditions often multiply rapidly until they begin to exhaust their food supply, and eventually a state of equilibrium is reached where prey and predator populations remain fairly constant. This is achieved by the complex interwoven food chains which ensure that any species is limited to a given population either by being eaten or by having a restricted supply of food.

I mentioned before that a large collection of freshwater invertebrates can easily be obtained, and these can provide hours of interesting and educational entertainment; with the help of a small aquarium a 'mini pond' can easily be set up. Rain water should be used, or water from the pond or stream being studied. Most of the more common animals can be caught using a small net – sifting through the dense vegetation and stones and sorting the catch in a white dish of some sort. Do remember though, that many of the creatures you will find feed on other invertebrates; if the numbers of predators are kept to a minimum it should be possible to set up a stable community. The only feeding required should be dead vegetation and detritus and the occasional 'top up' of some of the more heavily preyed-on species.

In a natural environment it is usually found that those which are fed on the most are the most abundant types. For example, in a stream I once studied in Wales the most abundant species were the larvae of several flies. These were being eaten by no fewer than a dozen other species, including fish which accounted for the majority.

Thus, if setting up a 'mini pond' it must be remembered that only one or two of the predacious invertebrates should be included. Obviously shrimps, pond lice, water fleas, lesser water boatman, water mites and the host of mayfly larvae, caddis fly larvae and mosquito larvae, etc. can safely be housed together. Others which can be included in small numbers are crayfish, leeches, flatworms, *Notonecta* and some of the carnivorous beetles.

With this variety of shapes and colours a collection of freshwater invertebrates can be almost as entertaining as an aquarium full of tropical fish. The only disadvantage is the lack of size of most of the species concerned. One important point – any such aquarium must be covered securely since many of the species concerned can and will fly!

It is interesting to consider how animals adapt to the environment, or should I say, choose the environment they are best suited to. This is demonstrated very well by several varieties of freshwater life: a good example is the limpets – their shells are shaped to suit the environment in which they live: the lake limpet has a rounded shell whereas the river limpet has a much more streamlined, hooked shell which offers less resistance to fast flowing water – a condition the lake limpet does not have to contend with. Another good example is the freshwater shrimp *Gammarus* sp. Its flattened body enables it to hide in tiny crevices and among weeds, etc. and consequently it can survive being attacked by fish and other animals simply by hiding out of reach.

One can easily go on with various examples but studying adaptations to an environment can be another fascinating way of learning how the ecology of a body of water holds together. Of course, it should be explained that the animals themselves do not, as such, adapt to the environment. It is merely a case of the survival of those species best suited to a particular ecological niche. For example, let us consider the survival of an imaginary 'new' animal. For various reasons it can only survive in very hot conditions, so it populates tropical jungles and hot springs all over the world, but gradually, as the competition for food increases in the tropical jungles those populations die out and only those living in the hot spring survive. This is, of course, a very extreme example; under natural conditions many more factors are responsible for the eventual make up of the fauna and flora of a given environment, but the same principles apply to our freshwater invertebrates, i.e. those best suited to survive in a given environment will do so, often at the expense of animals less prolific or less suited to that environment.

Figures in red beside a picture show the scale of reproduction

FOOD WEB

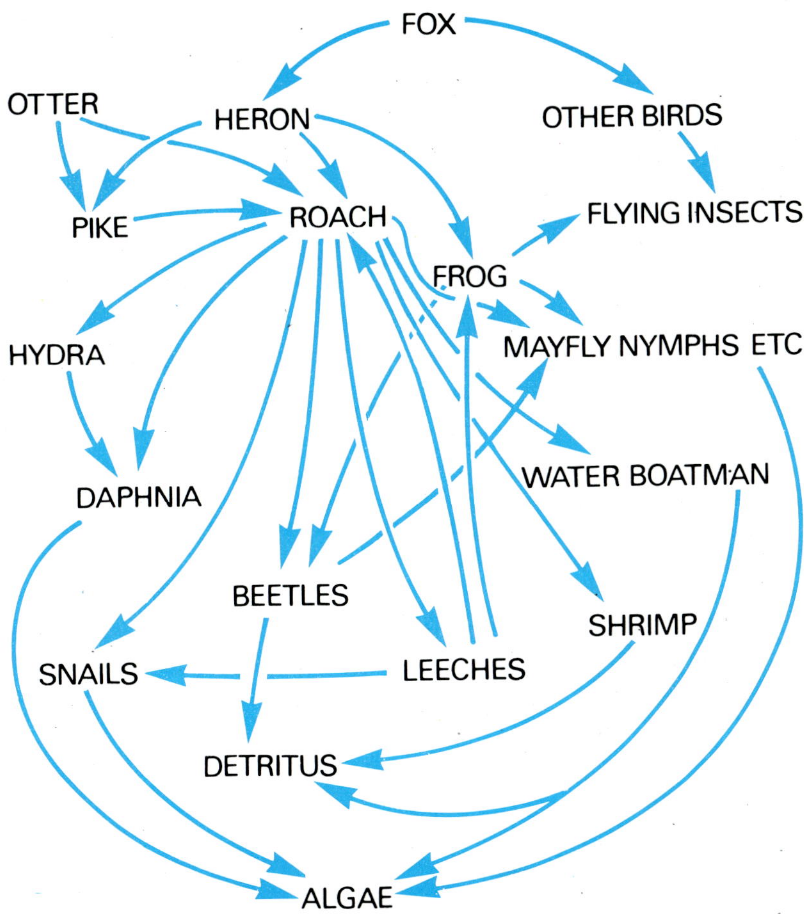

This diagram shows a very simple food web which is, in effect, a series of interwoven food chains. This example is quite typical of the situation often found in and around a body of water although, of course, in reality many more species are involved and the number of predator/prey relations is far greater.

To conclude I would like to stress one very important point. Some of the animals you may find will occur in very small numbers and unless these are required for some special reason, they should be returned unharmed. In these days of conservation it is easy to see how indiscriminate removal of rare specimens can lead a species to extinction, a fate we must all avoid!

1 *White-Spot infecting a Stickleback* ×2

2 ×10

White-spot is a well-known parasite which is often mistaken for some kind of fungal infection. It is in fact, a single-celled animal known as *Ichthyophthiriasis*.

Heavy infections of this protozoan parasite can cause serious depletions of fish populations with the result that the invertebrates normally eaten by the fish rapidly increase in numbers and can bring about a very unbalanced situation. However, it is usually only sick or injured fish which are weak enough to become infected, and consequently, under normal conditions 'Ich' merely tidies up undesirable injured fish.

The hydra is in the same group of animals as the jellyfish and sea anemones and occurs in Britain in two colour varieties – green *Hydra viridis* and brown *H. fusca*. They occur in weedy, still to slow-moving waters abundant in planktonic life. They can be seen as tiny green/brown cylindrical animals with up to eight tentacles extended ready to trap any passing particles of food.

Specialised cells found all over the tentacles entangle and kill the prey, which is then drawn towards the central column and forced into the 'gut'. Hydra are capable of movements by a

1. **WHITE-SPOT** *Ichthyophthiriasis*
2. **GREEN HYDRA** *Hydra viridis*
3. **FLATWORM** *Dugesia* sp.
4. **FLATWORM** *Polycelis* sp.
5. **FLATWORM** *Dendrocoelium lacteum*

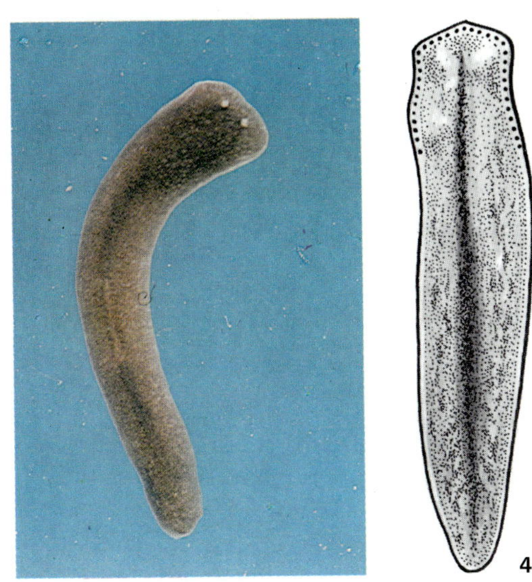

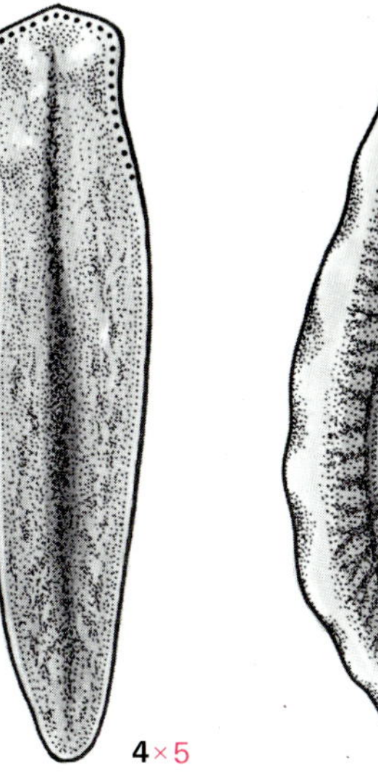

4 × 5

5 × 3

3 × 5

head-over-heels action but for only very short distances. Some reports have stated that a kind of swimming action has been observed.

Occasionally 'buds' can be seen on the body which may develop directly into small hydra, or the egg stage, which is capable of withstanding drought and freezing conditions.

The flatworms are some of the most primitive animals to be found in fresh water. Although they resemble small slugs they are by no means as advanced evolutionarily, as their molluscan counterparts. They move in a gliding motion by the use of millions of tiny hair-like projections, called cilia, along the underside. They have an incredible power of regeneration and if a flatworm is cut in half, two flatworms will eventually grow.

The three common species – shown here – are *Dugesia* and *Polycelis*, both dark brown to black, and *Dendrocoelium*, a beautiful white animal which can grow quite large. They can be found almost anywhere in both still and running water although *Dendrocoelium* prefers streams.

Small eyespots can be seen in many species but it is doubtful whether these are any more than sensitive to light or dark. They are mostly scavengers feeding on dead animal material, but a few species are capable of capturing small prey.

6. WORM *Lumbriculus variegatus*
7. WORM *Tubifex* sp.
8. LEECH *Erpobdella octoculata*

6 ×4

7 ×5

The true worms which inhabit our fresh waters include *Lumbriculus* and *Tubifex*. Both have a very similar appearance and life style but when examined carefully are quite separate in their characteristics. *Lumbriculus* is a much shorter, thicker animal and has very few small hairs, hardly visible, even with help of a lens. *Tubifex*, on the other hand, is an extremely long, thin worm with the front half having a generous supply of hair-like projections known as setae. Both live in burrows in the mud with the anterior end buried.

Tubifex can often be found, if conditions are ideal, in large 'beds' containing literally thousands of individuals. They feed on suspended particles and detritus found in the mud.

Of the thirteen species of leeches found in the fresh waters of Britain *Piscicola* is the most distinctive type. It is parasitic and attaches itself to various fishes and sucks blood from its host.

Sick fish are usually more prone to this parasite than fit ones, although *Piscicola* is capable of swimming at a substantial speed and can occasionally be found firmly attached to very large healthy fish. They grow to a size of about 10 cm and, as can be seen from the illustration, the suckers are very large compared to the long, thin body which tends to be quite firm.

They are fairly common and frequent still waters, especially large lakes, although they are seldom found in large numbers. The colour seems to vary from an olive-green through to a dull chocolate-brown.

9. LEECH *Glossiphonia* sp.

10. LEECH *Piscicola geometra*

11. LEECH *Helobdella stagnalis*

8 ×1

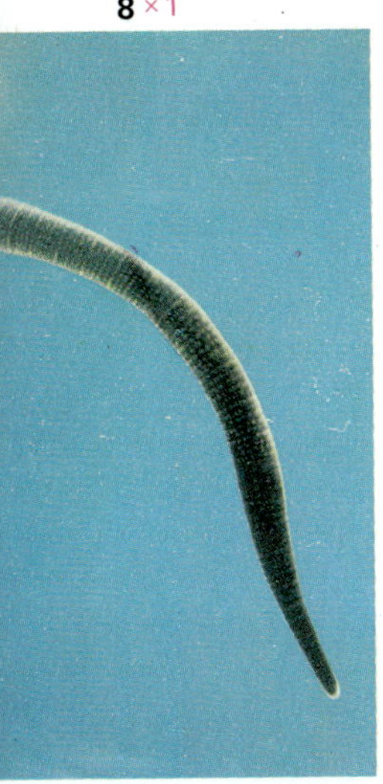

Glossiphonia, another common leech, is found mainly in running water attached to stones. It feeds mainly on other invertebrates such as snails, worms and insect larvae, and when disturbed will often almost roll into a ball. The colour varies from green to brown and this species does not swim.

Erpobdella is perhaps the most common of the British leeches; it is soft bodied and able to roll itself up. It can usually be found on, or under stones or attached to plants in both still and running water, and large numbers can often be found in particular ponds or areas. One case which comes to mind is a small lake in my area where up to fifty specimens have been found in a sample square metre of a weed bed. I understand, however, that populations of this density are unusual. They feed on various other invertebrates and are usually the dark reddy brown colour illustrated, although the light spots are not always visible.

Helobdella has very similar feeding habits but is found mainly in slow-flowing rivers and lakes. The body is almost transparent and the gut system is easily visible.

9 ×1·5

10 ×2

11 ×1

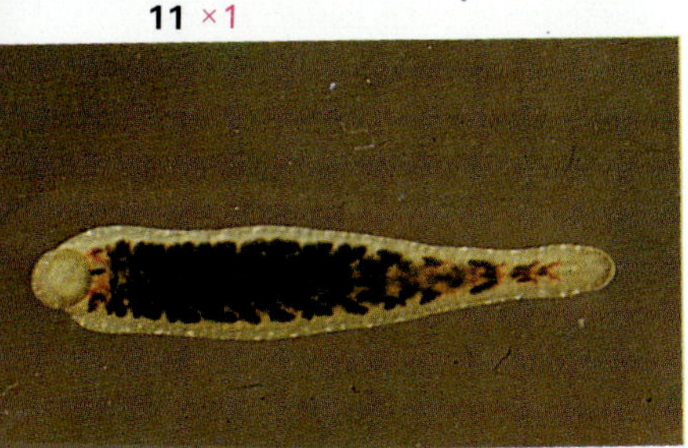

12. WATER FLEA *Daphnia* sp.
13. WATER FLEA *Cyclops* sp.
14. WATER FLEA *Candona* sp.

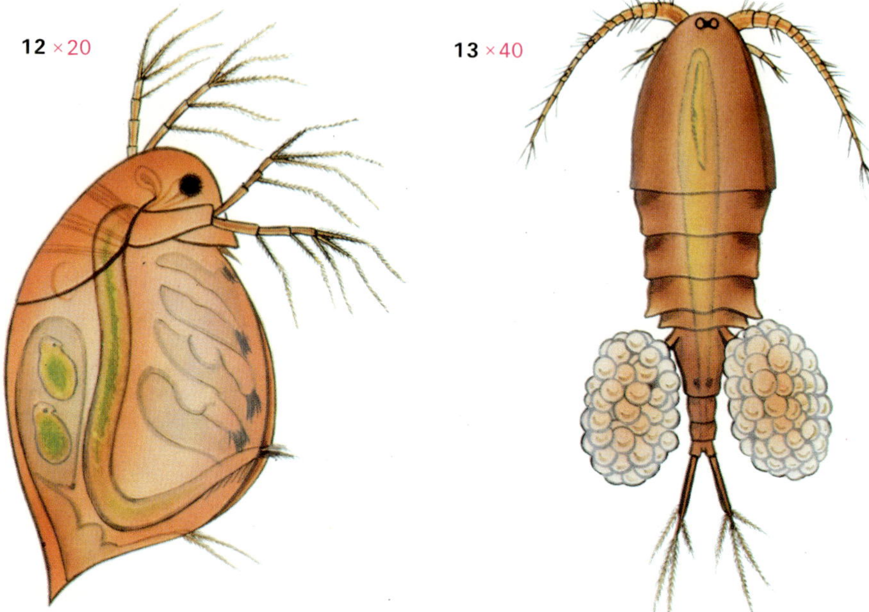

The Arthropods (meaning jointed limbs) make up a large proportion of the invertebrate world and include everything from water fleas to lobsters and the host of insects which inhabit the world. One of the most primitive groups of the Arthropods is the Crustacea. These are very well represented in the fresh water of Britain, the three most primitive groups being the Cladocera (water fleas), Copepoda (*Cyclops*) and the Ostracoda (*Candona*).

Daphnia the water flea, is not a flea at all. These small, but fascinating, and very important crustaceans occur in nearly all bodies of still water and in some of the slow-flowing rivers. In some areas several species may occur together but in certain ponds, where ideal conditions prevail, a single species may be found swarming by the million at certain times of the year.

In the early spring, eggs, which have been lying dormant through the winter, hatch, producing an all-female population. Throughout the summer and autumn these females produce more females without the need for fertilisation by males. Only as winter approaches are a few males produced which fertilise the next generation of 'dormant' eggs. These eggs can resist both drought and hard frosts and will only hatch as the water temperature increases. Thus populations of these tiny animals vary tremendously throughout the seasons. In summer they feed mainly on bacteria and suspended algae and provide a valuable food source for fish and other invertebrates.

They move by short jerking strokes of the long, branched antennae. Other limbs, protected by the shell-like body, collect food and also act as gills. Their colour can vary, depending upon what they have been feeding on, from reddy brown to a bright green if the gut is full of algae.

The characteristic shape of the Copepods is well illustrated by *Cyclops* sp. This is much smaller than *Daphnia* and has a distinct body and tail, the latter of which often carries two large egg sacs. The colour varies from dark olive-green to a reddish brown, and like *Daphnia* the antennae, which are not branched, are used for locomotion. Indeed, the very jerky movement of these creatures conjures up the idea of fleas much more so than does *Daphnia's* movement. They are usually found together with *Daphnia* although in much smaller numbers. One seldom finds water containing only *Cyclops* sp. Unlike *Daphnia* only eggs are produced in the reproductive cycle, some being quick developing and others resistant, dormant eggs. Both give rise to minute larvae which after about twelve months achieve adulthood.

The typical shape of the Ostracods is shown by *Candona* – a french bean would be the nearest analogy. About 230 species of these live in the fresh waters of Britain, for obvious reasons I shall not discuss them all!

Once again the antennae are used for swimming, although some species merely crawl among vegetation or burrow in soft mud. They feed mainly on dead vegetation, detritus and dead animals. Their reproductive cycle is somewhat of a mystery but from what is known it seems to be very similar to that of *Daphnia*. Both the eggs and adult stages can survive drought and freezing.

They tend to occur in weedy still waters and only in relatively small numbers, although I have, on one occasion, found that these outnumbered both the *Daphnia* and *Cyclops* in the suspended plankton of a local gravel pit.

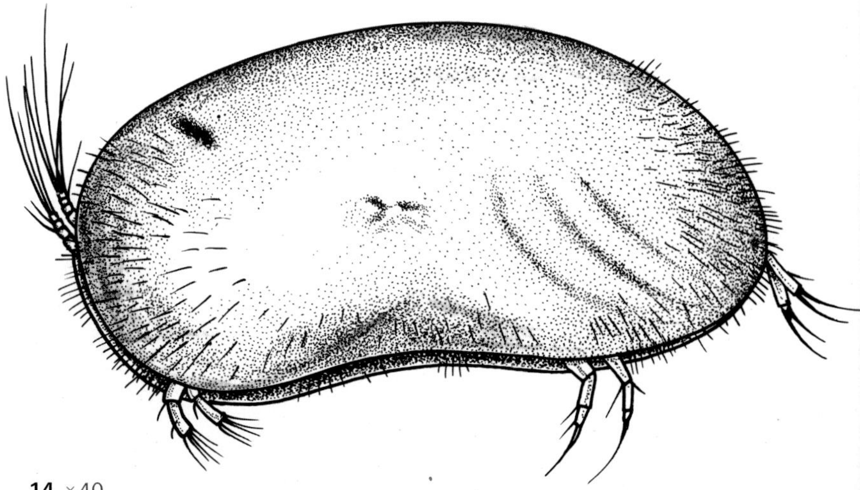

14 ×40

Although completely soft bodied, the fish louse is in fact a crustacean like the shrimp. It is superbly adapted for its parasitic form of life by being firstly almost transparent and secondly flattened to such an extent that it offers no more resistance to a flow of water than a normal fish's scale. They appear to attach themselves to parts of the fish which are provided with a good flow of water, such as the head, fins, around the base of the tail and just behind the gills.

Plagues of these parasites have been known to almost wipe out a complete population of fish up to very large sizes, the sheer number and weight of the parasites slow the fish down, and as the protective body slime is eaten so infection sets in and the fish dies. *Argulus* will continue to feed once its host is dead, but once all useful food has been removed, the fish louse will simply swim off in search of a new host. Under normal circumstances they occur in both running and still water in small numbers and one can occasionally find fish, surviving quite happily with twenty or more of these parasites covering the body.

Another crustacean, the pond louse, is slightly smaller than the freshwater shrimp *Gammarus*, the largest specimens being slightly less than 12·5 mm in length. They differ from the shrimp in being dorsally compressed and are less capable of swimming movements. They spend most of their lives crawling along the bottom and among vegetation, in both still and running water and play a similar role to the shrimp in breaking up dead vegetation and assisting in its decay. The colour seems to vary less than the shrimp and is almost always black, although I have seen some specimens which were a pale violet-grey. There is another species almost white in colour which occurs in underground streams and caves, *Asellus cavaticus*. Because of their more

15 ×7

16 ×4

15. FISH LOUSE *Argulus foliaceus*
16. POND LOUSE *Asellus aquaticus*
17. SHRIMP *Gammarus pulex*

17 ×4

lethargic habits they are able to survive in quite low oxygen levels but appear to be less prolific than the shrimp, occurring in smaller numbers but in a wider variety of locations.

Gammarus is a small but very prolific shrimp which can be found in most waters, but especially in fast running streams where there is plenty of vegetation or gravel. They vary in colour from a dull brown, through olive-green to a bright golden colour, as shown in the illustration. They can grow to a size of 18 mm and given the conditions will breed rapidly. A small spring pond I once studied which covered no more than 1·3 m^2, during ideal conditions, produced over 100 shrimps a day! They can swim very rapidly for their size and being laterally compressed are beautifully adapted for hiding in tiny crevices between stones. They feed mainly on detritus, i.e. dead vegetation, and play an important part in the initial breakdown of vegetable matter. Their breeding habits are very similar to those of other crustaceans – the female carrying the eggs for some time before they hatch, and these can often be seen in the spring and early summer.

18. CRAYFISH *Astacus pallipes*

18 ×1

The crayfish is the largest of Britain's freshwater invertebrates, growing to a total length of over 12·5 cm. It occurs in specific locations and is often limited to particular rivers or even a specific stretch of a river. They are seldom found in still water, except where they have been introduced – but muddy conditions are damaging to the delicate gills and thus the crayfish is usually found in clear running water where there is a high oxygen and calcium level. The colour varies from a pale green to a reddy brown, and, as can be seen from the illustrations, they have the appearance of a small lobster. The claws, or cheli as they are correctly called, are mainly for defence and mating purposes but are occasionally used to catch wounded or dying animals such as small fish, worms and other soft-bodied creatures.

The male tends to be much larger overall although specifically the cheli are relatively larger and the abdomen, or tail, much narrower than that of the females. Mating occurs in late autumn when the male, using his cheli, rolls the female over and deposits a packet of sperms on the underside of the abdomen. The eggs develop here, attached to the female, for up to six months. The young grow rapidly to begin with and often remain attached to the female until they are nearly 25 mm in length.

Although seemingly slow moving, crayfish can swim backwards very rapidly by repeated flicking forward of the tail, enabling them to swim against very strong currents. They are usually found under large stone or in thick weed beds where they hide by day; feeding occurs at night.

Damaged specimens can often be found but it is uncertain as to whether they fight among themselves or whether they are attacked by predators, Legs and cheli will grow again if broken off and it is not unusual to find a specimen with one cheli longer than the other.

Crayfish, like most hard-bodied invertebrates, have to shed their hard skin in order to grow. During the first four years this process may occur three times a year but after maturity is reached, at a size of about 63 mm only one moult occurs each year. Immediately after the moulting process the old skin is often eaten but the new skin is very soft at this time and does not harden for several hours, making the animal very vulnerable, and consequently it normally hides in crevices until the danger is past. The skin is normally shed in one piece and this involves some incredible acrobatics and great feats of strength. Often a claw or leg will be lost in the process which, when it is completed, allows the animal to grow by only 2–5 mm, depending on its original size.

A brief study I made on crayfish during my later school days indicated that some larger specimens could be as old as eight to ten years, but less than 1 percent of a population is likely to be this old.

18a Female ×1

18b Male ×1

The water spider tends to inhabit peat cuttings and ditches but has also been found in other types of still and slow-moving water. It is the only spider which spends its entire life submerged. Being a true spider, unlike the water-mites, it is capable of spinning a web underwater! Various situations are chosen for the web but usually it is made between leaves or aquatic plants and takes the form of a bell-shaped tube. Several of these may be made by one individual and each may be used for feeding, resting and mating.

Air bubbles are taken from the surface down to the webs which are filled with large pockets of air. The breeding nest is usually constructed close to the surface and camouflaged with various bits of vegetation. It consists of two chambers. In the upper chamber the eggs, up to 100, are laid, and the female occupies the lower chamber, from where she guards her brood and supplies it with fresh air bubbles from time to time. The young hatch after two to three weeks and remain in the nest for some time.

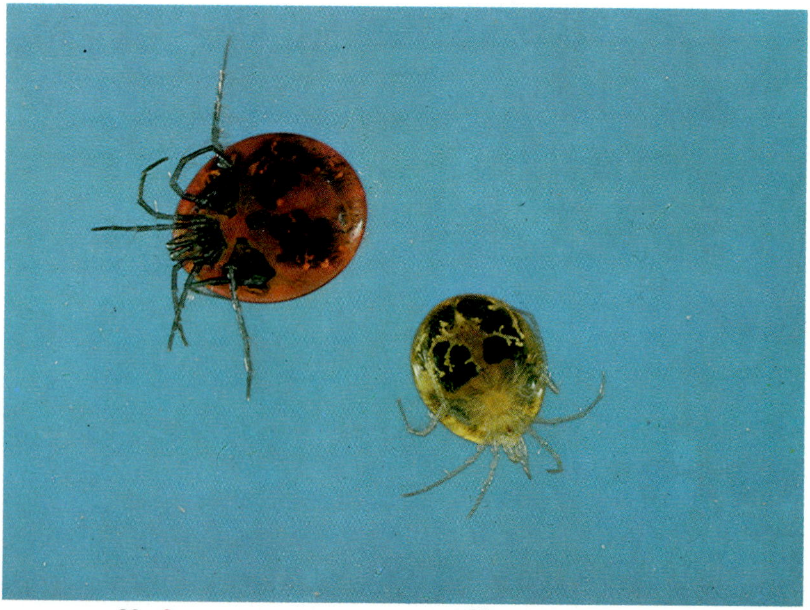

19 ×5

20 ×6 **21** ×6

19. WATER SPIDER *Argyroneta* sp.
20. WATER MITE *Limnesia* sp.
21. WATER MITE *Hygrobates* sp.
22. WATER MITE *Neumania* sp.

22 ×8

The diet consists mainly of small crustaceans such as *Daphnia* and larvae of various midges and flies. Special winter shelters are built by the adults but it has been reported that younger spiders will occupy dead snail shells which they seal off and fill with air so that they float to the surface.

The water-mites are fascinating and colourful spiders apparently evolved from the terrestrial spiders (such as we see spinning webs). They occur almost everywhere where there is a permanent body of water. They are predators, attacking minute water fleas, etc., using their sucking mouth parts to extract the liquid content. The life history is a very interesting one involving alternate parasitic, and free-living larval and pupal stages. In all there are six stages between egg and mature adult. A quick glance at No. 24, the pond skater, will show some of these parasitic stages as small reddish blobs attached to the outside of the body.

Unlike the water spider *Argyroneta* sp. these spiders have a breathing system which does not require them to come to the surface.

The species found in Britain can often be identified by the colour and patterning which cover a very wide range as shown in the illustrations. Their size is usually not more than 1–2 mm but some species grow to about 4 mm.

There are two specific types: the larger, short legged varieties which tend to swim a lot; and those with longer hairy legs, which often crawl among vegetation.

23. *Plea leachi*
24. POND SKATER *Gerris* sp.
25. *Microvelia* sp.
26. WATER STICK INSECT *Ranatra linearis*
27. WATER SCORPION *Nepa cinerea*

Plea leachi is a tiny insect often less than 3 mm in length. It is actually related to *Notonecta*, the water boatman, and behaves in much the same way as its larger relative. Indeed it could easily be mistaken for an early nymph stage of *Notonecta*. However, the much broader head and distinct wing covers are its distinguishing marks. *Plea* is usually found in dense vegetation at the water's edge and seldom occurs in large numbers except in southern England.

Gerris, or the pond skater, is a common insect found mainly in still water in sheltered bays or corners. It spends most of its life 'skating', as its name suggests, on the surface of the water by use of its very large and fragile legs. It can, however, dive and survive under water for some time, should the need arise. The very much shorter front legs are used mainly for catching food, both above and below the surface. The colour varies from medium brown to black and some species grow to a size of 15 mm.

Microvelia, like the pond skater, is a creature which relies on surface tension. It feeds mainly on insects which fall on to the surface, and may also catch mosquito larvae from under the surface film. They can be found in similar surroundings to the pond skater but are substantially smaller and more difficult to see.

24 × 1

The small red blobs are larvae of water mites.

23 × 8

25 × 4

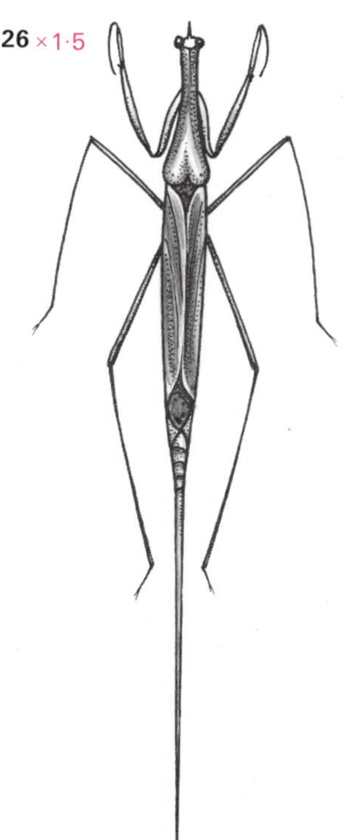

26 ×1·5

27 ×3

Ranatra, the water stick insect, is a large and predacious bug which tends to be localised in southern areas of Britain in still waters, dense in vegetation. Although similar to the water scorpion (*Nepa*) *Ranatra* is much elongated and is capable of dealing with much larger animals, including fish up to 40 mm in length. The very long breathing tube enables this insect to hang beneath the surface awaiting passing fish and other invertebrates which are caught using the front pair of legs in a similar manner to that used by the praying mantis.

The total length seldom exceeds 50 mm but I have seen specimens of 60 mm in the area around Reading (Berkshire).

The water scorpion, a strange, but superbly adapted insect, occurs widely throughout Britain in shallow still waters, where there is thick vegetation. It is unusual to find any number of these bugs in any given area. The extended breathing tube allows the water scorpion to suspend itself just below the surface with its front legs extended ready to trap any passing prey. A wide variety of animals are eaten, including tadpoles, small fishes and other aquatic insects.

The colour is nearly always a rich dark brown and they can grow to a size of 20–24 mm total length. The piercing mouth parts can easily be seen in the illustration as the long pointed proboscis extending forward from between the eyes and this is used to suck up the liquid remains of their prey.

28. LESSER WATER BOATMAN *Corixa* sp.

29. WATER BOATMAN *Notonecta* sp.

28 ×4·5

29 ×3

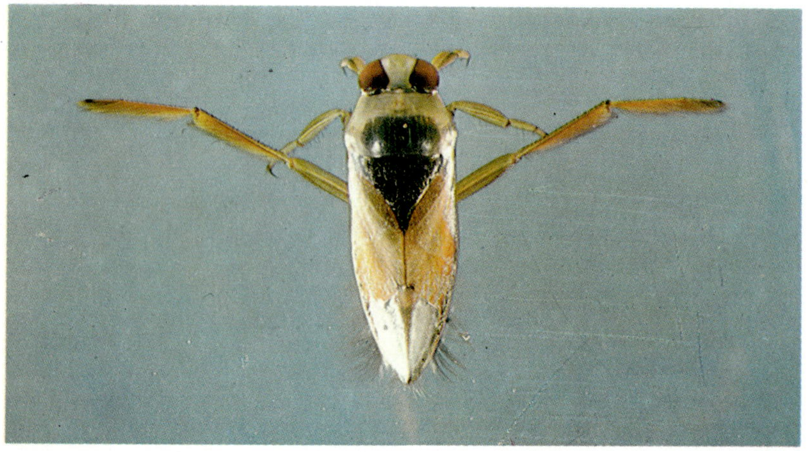

Notonecta, a predacious bug, can be found in almost any shallow, running or still water and is easily distinguished by its pink wing covers and large red eyes. Although both the adult and nymphal stages, as illustrated, do not show this striking coloration. Unlike the more common water boatman (*Corixa*), *Notonecta* rests at the surface upside-down, i.e. pink wing covers facing downwards, thus, as can be seen from the third illustration, a very dark brown body is presented to any would-be predators from above. *Notonecta* feeds on tadpoles, small fish and other invertebrates and an interesting point is that its swimming position is controlled by light, so that if a *Notonecta* is placed in a tank lit from underneath it will swim upside-down; or right side up – depending which way you look at it.

Both *Corixa* and *Notonecta* are capable of flying for substantial distances and I have once observed, on a hot summer's day, both *Corixa* and *Notonecta* moving from water to air and back again as if there were no barrier whatsoever. The reaction time must be truly amazing for a creature such as *Notonecta* to be able to open its wings at the instant of reaching and leaving the surface of the water and to close them with equal skill on the return dive.

Corixa, the water boatman, feeds mainly on decaying vegetation and is therefore usually found in waters thick with plant life. They can occur in enormous numbers and in one particular lake I have studied there have been counts of up to 250 per square metre! This, of course, was an exceptional case and the highest counts occurred in the shallower areas around the edges of the lake.

It has been reported that the male *Corixa* can produce a mating song by rubbing his hairy front legs against a ridge on the side of the head.

Both *Corixa* and *Notonecta* tend to occupy the shallower areas of water, for the simple reason that they both have to come to the surface to breathe. *Corixa* can stay under water for up to 5–6 minutes, depending on how active it is and also on the temperature of the water. *Notonecta*, on the other hand, tends to spend much more time hanging at the surface in its upside-down position awaiting its prey.

29a ×1·5

29b *Nymph* ×2

Although similar to *Gerris, Hydrometra*, the water measurer, is much smaller and less common except in localised areas where they can be found in small numbers. Their habits are similar to the pond skater although underwater activities are less, and more time is spent on the surface nearer the bank with the body raised high above the water.

They feed on small animals which come to the surface to breathe and on drifting dead animals. the elongated head is thought to help with the feeding, which involves actually piercing the surface film and grasping the prey under water.

Two species occur in Britain, one much more common, *H. stagnorum*, and one very rare, reported to exist only in the Norfolk Broads area, *H. gracilenta*. The one illustrated is *H. stagnorum* and as can be seen the head in front of the eyes is twice as long as that behind it, as opposed to the other species where this part is one and a half times as long. The colour varies from chocolate-brown to black and the maximum size is about 12 mm.

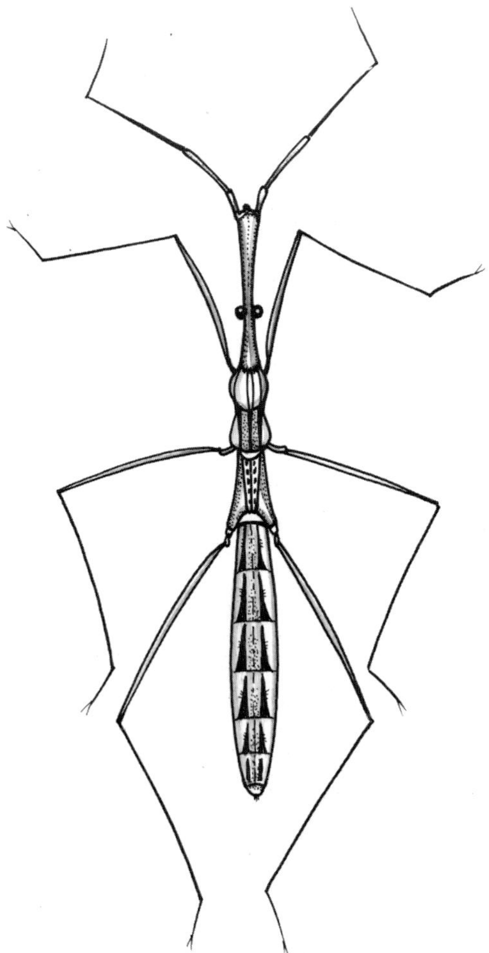

30 ×10

30. WATER GNAT *Hydrometra stagnorum*

31. *Helmis* sp.

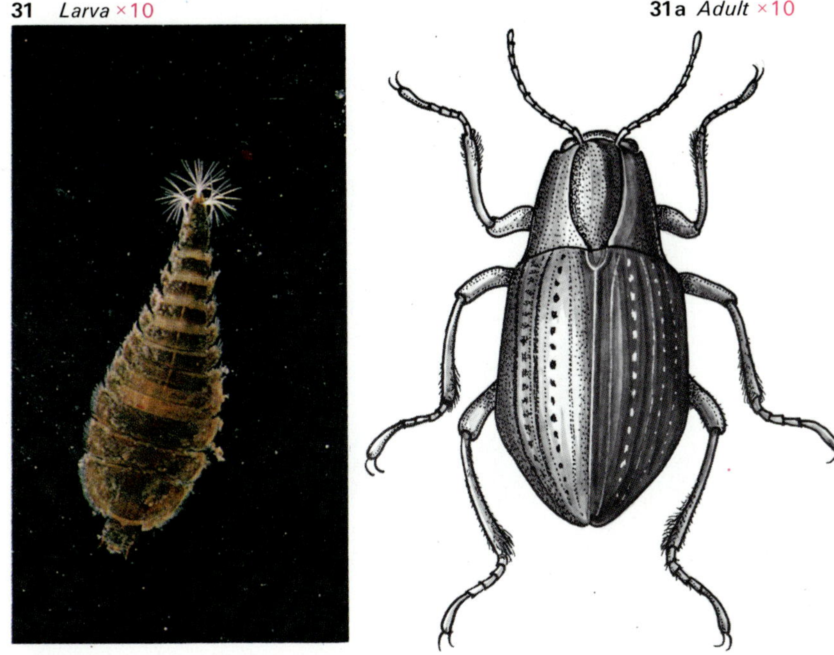

31 *Larva* ×10

31a *Adult* ×10

Having now dealt with the bugs we come to the more familiar beetles. True beetles have completely hardened wing covers and generally are much larger than the bugs. Only a few common species can be mentioned here but suffice to say that there are several hundred species common in Britain's fresh waters. It is important to remember that beetles have larval stages, and quite often the diets of both the larva and adult are virtually identical.

The first of the beetles very common in fast-moving streams is *Helmis* sp. Both the larva and adult live among stones and vegetation at the bottom of the stream where they graze on vegetation and detritus. The adult is not capable of swimming and spends its entire life crawling around the bottom, at a noticeably slow pace. They are usually jet black in colour and seldom more than 3–4 mm in length. The larval stage has three feathery appendages on its tail which are actually gills. Some streams will contain very large numbers of these beetles, for example, a mountain stream on the Welsh border I once visited produced counts of up to 100 individuals per square metre, but the distribution was confined to the fast-flowing stretches where the bottom consisted largely of small stones and debris. Sandy and muddy areas produced only one or two of these sluggish, long-clawed beetles.

32. GREAT DIVING BEETLE *Dytiscus marginalis*

33. WHIRLIGIG BEETLE *Gyrinus* sp.

32 *Larva* ×3

32a *Adult* ×1·2

33 ×4

The sheer size of *Dytiscus*, the great diving beetle, accounts for its common name, and it is strictly carnivorous, feeding on fish, tadpoles and other invertebrates. The wing covers of the female are distinctly marked with longitudinal ridges whereas those of the male are completely smooth.

The larva, also illustrated, is equally predacious and can handle fish, etc. of its own size (up to 35 mm). The mouth parts are superbly adapted for catching the prey; as can be seen the hooked shape of the jaws enables the larva to grasp its prey very firmly. Because the larva lacks the powerful swimming legs of the adult stage it is much less active, relying on crawling among vegetation although it can swim for short distances by somewhat frantic movements of all six legs. The feeding position of both larva and adult is hanging at the surface but the adult often hunts down its prey under water and has been observed actually chasing small fish.

Another very active beetle, the whirligig, can often be seen in large swarms in the margins of still waters or in quieter stretches of running water. The shiny black wing covers are characteristic of the species and make them easy to see as they skate around on the surface of the water. (They are capable of diving but tend to do so only when disturbed.) An interesting point about the whirligigs, of which there are twelve species in Britain, is that their eyes are adapted for vision both in and out of water. The upper half is anatomically adapted for aerial vision whereas the lower half is adapted for vision under water. Thus, when resting – laying flat on the surface – they can see equally well above them and below.

The front legs are 'normal' and are used for catching the prey, which consists of living and dead animals floating at the surface. The middle and hind legs are short, paddle-like limbs superbly adapted for swimming (the sheer speed and manœuvrability of these beetles only goes to prove this point).

34. ***Hydrobius*** **sp.**
35. ***Haliplus*** **sp.**
36. ***Hygrotus*** **sp.**

34 ×3·5

35 ×8

36 ×8

Hydrobius is another of the more common beetles which is usually found in small numbers in stagnant ponds. (At least, it has seldom been reported in running water.)

The adults are largely vegetarian, spending their time crawling among and grazing on decaying vegetation. Some of the larval stages are, however, carnivorous. A distinguishing factor of this group of beetles, the Hydrophilidae, is that the underside of the body is covered in stiff hairs which trap air, which is taken from the surface.

Haliplus sp. is a crawling beetle, although its more hairy legs do enable it to swim for short distances. They actually prefer crawling among the filamentous algae on which they feed. The distinctive coloration and pattern of the genus makes them fairly easy to identify, and they appear to be fairly common throughout England and southern Scotland. They are usually very small and seldom exceed 3 or 4 mm in length.

Hygrotus, an attractive and very distinctive beetle, is fairly common in still waters throughout Britain and I have occasionally found specimens in the deeper, slower-moving stretches of small streams, especially in Norfolk. It is quite a strong swimmer and feeds mainly on smaller invertebrates such as flatworms, *Daphnia*, etc. The actual patterning of the body colours appears to vary somewhat but the example shown is fairly typical. The size seldom exceeds 4 mm.

The lake limpet, *Ancylus lacustris*, is one of the smaller Gastropods living in Britain. It seldom exceeds 7 or 8 mm and is usually the mottled brown colour as illustrated. It grazes mainly on plants and detritus and can often be found in stagnant water. The

37. LAKE LIMPET *Ancylus lacustris*
38. RIVER LIMPET *Ancylastrum fluviatile*
39. ORB SHELL COCKLE *Sphaerium* sp.

shell is much larger than the soft body, which is thus well protected. It tends to be restricted to still waters where it does not have to compete for food with the river limpet.

Ancylastrum, the river limpet, is very closely related to *Ancylus* but is slightly larger and more conspicuous. The characteristic 'hooked' shell actually helps *Ancylastrum* to resist strong water currents, and consequently fast rivers and streams tend to be the favourite habitat. *Ancylastrum* grazes on stones and plants and quite often large numbers of individuals can be found in isolated pockets.

The orb-shell-cockle, *Sphaerium* sp., is in many ways similar to the edible cockle, except, of course, for its size. The shell is often almost transparent but usually displays the golden, yellow colour shown in the illustration. The two siphons, easily visible, are used both for passing water over the gills and for feeding. It is usually found in streams and slow-moving rivers with sandy or gravelly bottoms.

37 ×2·5

38 ×2·5

39 ×5

40. RIVER SNAIL *Viviparus* sp.

41. RAMSHORN SNAIL *Planorbis* sp.

42. EAR POND SNAIL *Limnaea (Radix) auricularia*

43. GREAT POND SNAIL *Limnaea stagnalis*

40 ×1·2

41 ×1·5

42 ×2

The river snail, *Viviparus*, an attractive snail, with its striped shell, can often be found in small groups of up to ten individuals. It tends to inhabit running water but has been reported in some still waters where there are plenty of large stones, under which it normally hides.

Like most of the freshwater snails its diet consists mainly of algae and detritus on which it grazes. The scientific name *Viviparus* means 'live bearing'. The young are reared in the uterus of the female and fed on a milky fluid until they are fully developed. They are about 10 mm long when they are born and the shell is covered by bristles which eventually fall off.

43 ×1

43a *Eggs of* Limnaea stagnalis ×1·5

Unlike other snails the sexes are separate: the male can be identified by its smaller right tentacle which tends to be thickly knobbed.

The ramshorn, *Planorbis* sp., is perhaps one of the more attractive snails; its beautifully whorled shell and bright colour make it easy to identify. Some species can grow to about 30 mm (the great ramshorn, *P. corneus*), but most are quite small. They inhabit both still and running water but tend to occur in larger numbers in large, deep lakes.

The ear-pond snail, *Limnaea auricularia*, is one of the less common of this genus and is normally found in still water dense in vegetation. The characteristic flattened, wide shell is typical of the species and is supposed to resemble an ear.

The great pond snail, *Limnaea stagnalis*, is very common in lakes with abundant vegetation in central and southern England. It seldom occurs in large numbers, although I have on occasions counted as many as a hundred feeding very close to the surface, within a twenty-metre stretch of a weed bed. The shell tends to be quite dark, and large specimens can often be found with substantial growths of algae covering the shell.

The eggs, as illustrated, are laid in large masses of jelly and, during the summer, develop very quickly. Many of the young, however, fall victim to numerous predators.

44 × 2·5

45 × 1

The glutinous snail, *Myxas glutinosa*, is very rare in some areas but is found in much of south-east England in still water with abundant vegetation. The beautiful, thin shell with its mottled pattern is characteristic of the species and although some other species display this mottling effect, none are quite as distinct as *Myxas*. It is generally found grazing on algae and detritus but may also occur on some of the larger leaves of aquatic plants.

Perhaps the most common species of our freshwater snails, the wandering snail, *Limnaea pereger*, inhabits most types of water and can often be found in larger numbers grazing on algae and detritus. It feeds, like most of the snails, by rasping particles of food from stones and weed using its horny tongue or 'radula' which is covered by rows of thick tooth-like projections.

It moves by regular contractions and relaxations of the muscles in the 'foot' and glides along on a track of mucus which is continually secreted from glands covering the base of the foot. This mucus track also enables *Limnaea* to move along upside-

44. GLUTINOUS SNAIL *Myxas glutinosa*

45. WANDERING SNAIL *Limnaea pereger*

46. MARSH SNAIL *Limnaea palustris*

down on the surface of the water, but if disturbed, air is expelled from the gill cavity and the snail sinks rapidly to the bottom.

Air, held in the gill cavity, can also enable these snails to rise, like corks, to the surface. The muscles of the cavity wall merely relax, allowing expansion of air inside, and on releasing the grip of the foot, the animal floats upwards.

Limnaea palustris, the marsh snail, is very widespread, though not abundant. It can be found in ponds and slow-moving rivers and may also be found in brackish and even salt water. The shell shape is rather similar to that of *L. stagnalis*, but the overall size is much smaller and the shell much thinner and mottled. It is usually found grazing on vegetation and can occur in very large numbers under good conditions.

Snails are often confined to waters with high calcium levels because this element is required for building the shell. Indeed, the lack of calcium in some waters can change the form of the shell quite markedly within one species.

Many species are carriers of various parasitic flukes or certain stages in the life history of these flukes. Indeed, the Limnaeidae are involved in two stages in the life history of *Fasciola hepatica*, the liver fluke which infects domestic animals and humans.

46 ×1·5

47. SWAN MUSSEL *Anodonta cygnea*

47 ×1

The swan mussel, *Anodonta*, is very similar to the edible mussel although the shell tends to be much lighter in colour and more rounded. The general colour is olive-green to brown and they can usually be found, half buried in mud or sand, in both running and still water. They are 'filter feeders', using their gills to filter out tiny particles of food from the water flowing over the respiratory surface.

The life cycle involves a larval stage called the glochidium larva which is released in thousands from the brood pouch, inside the gills. These larvae are parasitic and attach themselves to the fins of the host fish – by means of large claw-like projections of the shell. They remain here for several weeks before they are mature enough to lead a self-supporting life. Some species can release up to a million of these larvae at one time and consequently a water heavily populated by these molluscs will often hold fish carrying large numbers of larvae at certain times of the year. *Anodonta*, and related species, play an important part in the ecological system – some of the larger species can filter up to twenty gallons of water per hour: they extract food particles and mud and help to maintain a balance of the nutrient material suspended in the water.

They can often be found in large numbers – up to ten individuals per square metre is not unusual.